Busy Bugs

SPIDERS

By Bray Jacobson

Please visit our website, www.garethstevens.com. For a free color catalog of all our high-quality books, call toll free 1-800-542-2595 or fax 1-877-542-2596.

Library of Congress Cataloging-in-Publication Data
Names: Jacobson, Bray, author.
Title: Spiders / Bray Jacobson.
Description: New York : Gareth Stevens Publishing, [2022] | Series: Busy bugs | Includes index.
Identifiers: LCCN 2020008097 | ISBN 9781538263396 (library binding) | ISBN 9781538263372 (paperback) | ISBN 9781538263389 (6 Pack) | ISBN 9781538263402 (ebook)
Subjects: LCSH: Spiders–Juvenile literature.
Classification: LCC QL458.4 .J37 2022 | DDC 595.4/4–dc23
LC record available at https://lccn.loc.gov/2020008097

First Edition

Published in 2022 by
Gareth Stevens Publishing
111 East 14th Street, Suite 349
New York, NY 10003

Editor: Kristen Nelson
Designer: Katelyn E. Reynolds

Photo credits: Cover, p. 1 LionH/E+/Getty Images; p. 5 Alongkot Sumritjearapol/Moment/Getty Images; p. 7 sandra standbridge/Moment/Getty Images; p. 9 Andrea Incerti /500px/Getty Images; pp. 11, 24 (eggs) Janny2/ iStock / Getty Images Plus; p. 13 dennisvdw/ iStock / Getty Images Plus; pp. 15, 24 (shed) Astrid860/ iStock / Getty Images Plus; p. 17 Jose A. Bernat Bacete/Moment/Getty Images; p. 19 Chico Sanchez/Getty Images; pp. 21, 24 (web) Albert photo/Moment/Getty Images; p. 23 shikheigoh/ RooM/ Getty Images.

Printed in the United States of America

CPSIA compliance information: Batch #CSGS22: For further information contact Gareth Stevens, New York, New York at 1-800-542-2595.

Contents

Spiders live all over the world.

They can live in warm
or cold weather.
They live in water!

They have eight legs.
They have two body parts.

Mothers lay eggs.
There can be hundreds!

Babies look like tiny adults.

They shed skin as they grow.

Spiders live alone.

They move fast.
They hide!

Many make webs. Some webs help catch food.

Spiders eat bugs.
Some eat other spiders!

Words to Know

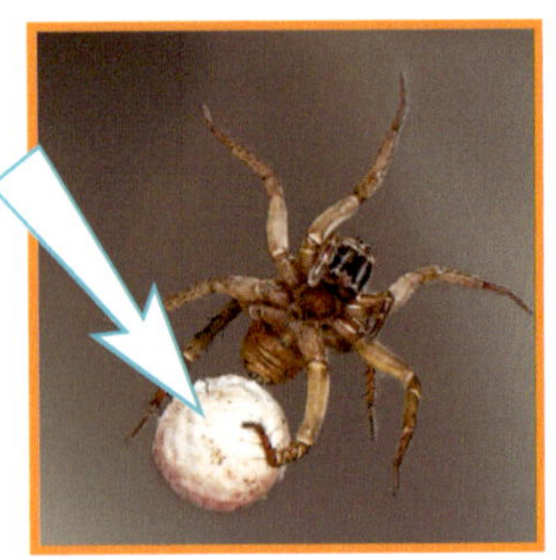
eggs

shed

web

Index